Smart Green Civilizations
Ancient Greece

teri
The Energy and Resources Institute

A note from Dr R K Pachauri

Human civilization in its race towards progress has at times ignored its adverse effects on nature. With every passing century, man has intensified his quest for a tomorrow better than today, and issues like environment-friendly living, usage of clean energy and preventing the harmful effects of chemicals on nature are becoming increasingly important. Contrary to popular belief, these can be tackled without compromising on our comforts. All we need to do is turn a few pages of history and relearn lessons that civilizations from various parts of the world have left behind.

This series provides a unique and interesting perspective of history from the eyes of an environmentalist. It highlights the environmental wisdom of ancient people. These books bring alive ancient civilizations and their simple, earth-friendly lifestyles—building bright and airy houses from mud bricks, using the sun's energy to heat homes, utilizing plants to make natural dyes, applying manure to grow crops, and many more such techniques.

Exploring the fascinating civilizations of the ancient world and bringing forth little known 'green lessons' from the past, I hope these books will ensure that young readers put to use the knowledge of yesteryears to lay the foundation for a prosperous future.

R K Pachauri
Director-General, TERI
Chairman, Intergovernmental Panel on Climate Change

Contents

TERI MEETS ZEUS	6
SMART CITIES	8
FROM BIRTH TO SIXTY	10
GREEK GASTRONOMY	12
ANCIENT APPAREL	14
MOVING AROUND	16
THANK YOU GODS FOR EVERYTHING	18
FITNESS FREAKS	20
ANCIENT SCIENTISTS	22
ART AND ARCHITECTURE	24
THE BATTLES THEY FOUGHT	26
WHY DID THEY DISAPPEAR?	28
GREEN LESSONS	30

Teri meets Zeus

Teri was startled by the clap of thunder. She wondered why thunder was always heard after lightning. Just then, another streak of lightning flashed across the sky.

Teri felt herself being lifted.

The bolt of lightning stopped at a towering pillar.

YOU HAVE BEEN CHOSEN TO TRAVEL ACROSS ANCIENT GREECE! ZEUS, KING OF GODS, WILL EXPLAIN ALL...

Greece lies in south-eastern Europe. It is home to one of the earliest civilizations, or human settlements, dating as far back as the Stone Age, when humans had started using stone tools. The best years, when most of the developments in art, architecture, and science took place, were from 2000 BC to 146 BC.

The ancient Greeks developed new ideas on cities, rulers, the arts, sports, and medicine. The earliest Greeks were called Minoans after their famous King Minos. The Myceneans, who were excellent soldiers, came next. Around 480 BC, the civilization had developed ideas on politics, philosophy, art, architecture, and science that influenced many other civilizations across Europe and even Asia.

▶ The ancient Greek civilization, divided into small city states, also included parts of today's Turkey and Bulgaria, regions of Italy, France, and North Africa.

The ancient Greeks worshipped Nature. They were the first to study natural things like water.

SMART GREEN CIVILIZATIONS

The Greeks were curious about everything. They studied different subjects. We learn a lot about them from the writings of historians like Herodotus, Thucydides, and Xenophon. The Greeks also made buildings, temples, statues, and pottery. Some of them stand even today. The influence of ancient Greece can be seen far and wide—from Europe to North Africa and some parts of Asia as well.

and NOW
Ancient Greeks believed we could understand the world around us by imitating Nature. Today, people are realizing the importance of living in harmony with Nature.

Smart cities

Welcome to Mount Olympus. I'm Zeus, god of sky and thunder.... Light travels faster than sound, so you see lightning before you hear thunder. Now let me show you around Greece. I know that's your next assignment.

I wish I'd brought my notebook!

The ancient Greeks lived in hundreds of city states like Athens, Sikyon, Olympia, Delphi, Sparta, Thebes, Argos, and Corinth. They respected Nature and tried to utilize natural and renewable sources of energy like the sun and wind.

The Greeks planned their cities after studying the seasons. That is why entire cities faced the south. Socrates, the scholar, wrote, 'In houses that look south, the sun enters the portico in winter.' This kept houses warm and saved burning wood, which was scarce.

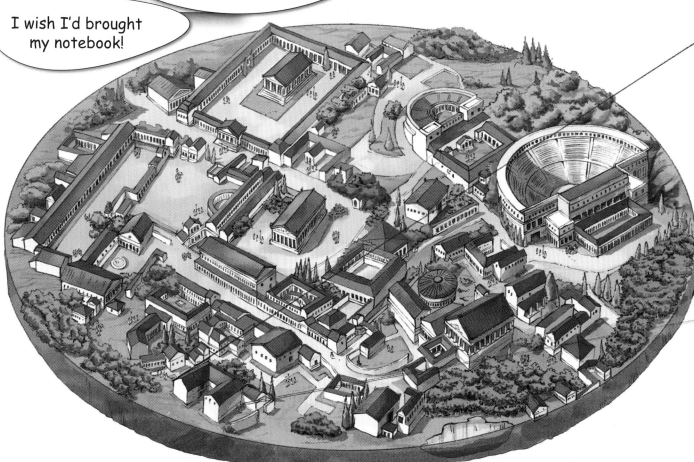

▲ Greek cities were well planned, with homes and public buildings like temples, open air theatres, stadiums, and gymnasiums.

 Ancient Greek plumbers used the force of water from aqueducts to build stone fountains, baths, and personal shower rooms with running water.

SMART GREEN CIVILIZATIONS

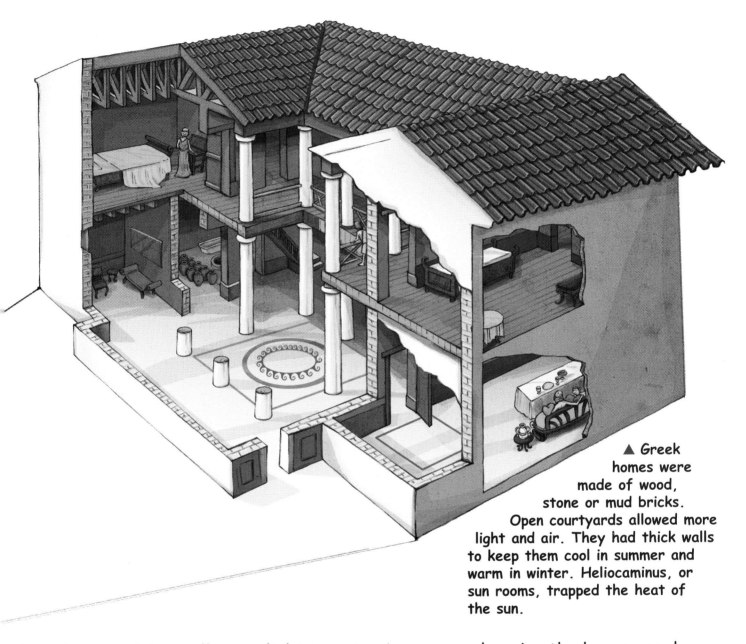

▲ Greek homes were made of wood, stone or mud bricks. Open courtyards allowed more light and air. They had thick walls to keep them cool in summer and warm in winter. Heliocaminus, or sun rooms, trapped the heat of the sun.

The designs did not allow sunlight to enter in summer, keeping the houses cool. Through the year, the houses were well lit.

The Greeks also studied wind movements and placed doors and windows in such a way that houses were airy. The north of the house was sheltered to keep out cold winter winds. Windows had marble shutters that could control the amount of breeze and sunlight that entered the house. This kept the house cool in summer and sunny in winter. The shutters could be bolted at night for safety or during storms, to keep rainwater out.

THEN and NOW

Most Greek buildings made good use of natural sunlight in their buildings. Today, some modern buildings use this technique, called passive solar design.

From birth to sixty

"Oof! Assignments! Did your children go to school? Do you think I'd have been happier among the ancient Greeks?"

"Girls didn't go to school. Now would you have liked that?"

When a baby was born, the father danced around the neighbourhood with the child, to announce the baby's arrival. The Spartans were very strict. Babies were checked by soldiers. Weak infants were taken away to become slaves. Each city state had its own way of teaching children.

▼ Apart from reading and writing, Greek teachers taught sports like wrestling, as well as music.

Instead of paper, students wrote on reusable wax-coated wooden boards, cutting the wax with a wooden stylus. Most lessons were memorized, rather than written, so they did not waste paper.

SMART GREEN CIVILIZATIONS

Education was expensive, so poor boys often could not afford to study. Athenian boys studied at home till they were seven years old. The few schools they had were run in the homes of the teachers. Girls stayed home and learnt housekeeping. However, girls in Sparta went to school and learnt wrestling and gymnastics. Boys found fit became soldiers at twenty years of age. At fourteen, children of tradesmen began to learn a trade.

Men ran the state and made sure their slaves worked in the fields. Some sailed to different lands to trade. They brought wheat from Italy, from Cyrenaica in modern Libya, and from lands around the Black Sea. Papyrus came from Egypt. Most men hunted. Some were fishermen who caught and sold fish. Some men spent their life studying science, the arts or philosophy. Some soldiers worked for other states, but most were soldiers for their own city states and served till the age of sixty.

▼ Greek children used a pointed pen to write on special wax-coated tablets.

▶ Most rich boys studied alone with a tutor or with a small group of boys of about twenty or less.

THEN and NOW
Only men could join the army, play sports or meet in the market. Today, most societies grant the same freedom to women.

Greek gastronomy

What an action-packed day! What did the Greeks eat to get so much energy?

We ate a lot of healthy food. Look, this is where we grew our food.

The Greeks were wise. They knew that you are as healthy as you eat, so the right food was important. They grew their own food and ate wheat and barley made into bread and cake. They were expert bakers and could make different kinds of bread like brown, rye, and milk bread.

The Greeks also ate fruit and vegetables like beans, asparagus, cabbage, and radish, and greens like celery. Sometimes, the fruit was baked. Food was cooked in olive oil. The grapes they grew were made into wine.

▶ Bakeries were always buzzing with activity, with bakers kneading dough and baking different varieties of bread.

▶ The Greeks ate fruits like olives, apples, grapes, figs, prunes, apricots, cherries, and dates. They also knew the art of making wine.

Olive oil is one of the healthiest oils. It is good for the heart and is rich in vitamins that protect body cells from damage. Fish and seafood fight infection and are good for health.

SMART GREEN CIVILIZATIONS

Herdsmen kept goats. They drank the milk and also made it into cheese. Since Greece is by the sea, the people ate fish and seafood like mullets and tuna. Most of the meat they ate was from small birds.

Greek farmers knew how to keep pests away from their crops. They used amurca, a bitter juice from crushed olives to keep away destructive insects and animals like moles. The juice also worked against weeds. Plant juices like hemlock and lupin, and animal dung mixed with water were sprinkled to keep insects away and kill larvae. Orchard farmers in Corinth covered leaves with the fat of animals like goats to keep insects away.

Wheat was planted in autumn, when it rained. Therefore, there was no need for irrigation.

and NOW
The ancient Greeks used methods of farming and pest control that were safe for the environment. They treated the soil with lime to make it more fertile. Modern farmers use chemical pesticides that cause pollution and damage our health.

Ancient apparel

All the walking around left Teri feeling hot and sweaty. But the heat didn't seem to bother Zeus.

My clothes are made of a natural fibre called linen. It allows my skin to breathe. Look at your thick trousers! You must be so uncomfortable!

Cool and comfortable!

For most of the year, Greece was a warm, dry place. So, people wore loose garments made of unstitched cloth, which kept them cool. They wore a knee-length tunic, or chiton. It was a large rectangular cloth held in place by pins at the shoulders and a belt round the waist. Women's dresses reached down to the ankles. Both men and women also wore a heavy cloak, or himation, over the chiton or by itself. Babies wore only cloth diapers in summer. Sometimes, they were not dressed in any clothes. As the weather cooled, they were swaddled, or wrapped in clothes.

▶ Clothes were woven at home by women and slaves and dyed using colours from vegetables.

The Greek hillside was once covered by trees. Around 800 BC, people started cutting down trees for firewood, so that they could burn fires to melt iron. Gradually, the hillsides became rather barren.

SMART GREEN CIVILIZATIONS

◀ The rich wore coloured clothes that flowed to the ground, while the poor wore simple clothes. In the house, they walked barefoot, but when they went out, women wore ankle-high shoes, and men wore strapped sandals.

▶ Men wore their hair short, while women kept it long. Women plaited the hair and then coiled the braid on top of the head.

In summer, people wore clothes made of linen. Winter clothing was made of wool. As Greek sailors began to travel further, they brought cotton from India, but it was expensive. Cloth was mostly woven at home by women and slaves. Dyeing cloth was not easy, and it was expensive, too. So, the poor wore plain tunics. Only the rich could afford to wear coloured outfits, dyed violet or green from natural colours made from plants.

and NOW
Clothes coloured with natural dyes did not cause any harm. Most clothes today are dyed using chemicals, which can be harmful to health and the environment.

Moving around

With thousands of miles of coastline and so many islands, how do you think the ancient Greeks moved about?

Much of Greece is rocky, so constructing roads and hardy vehicles was difficult. Greek traders moved goods inland down unpaved roads in wagons and two-wheeled carts. These were drawn by mules. Heavier goods were pulled by oxen.

The Greeks built strong boats of wooden planks. Instead of making a frame first, they joined planks of wood to make the body of the boat. Then, they fitted a frame into this. By the first century BC, a ship could carry four hundred tonnes of goods.

▼ The ancient Greeks were excellent boat builders. They made different types of boats—while the pentekonter had fifty oars in one row, the bireme had two rows of oars on each side.

 Most Greeks walked from one place to another. Some used mules rather than horses to carry goods and people. This way, they used little energy and no machines, so they did not cause pollution.

SMART GREEN CIVILIZATIONS

◀ Chariots drawn by two or four horses were a popular means of transport. Men went to marry in chariots!

The Greeks' excellent knowledge of astronomy helped them to navigate their boats. Sailors charted their course watching the stars for direction. The Greeks were also among the first to build lighthouses.

Ships were docked on land. The Isthmus (a strip of land between two seas) of Corinth had a diolkos, perhaps the first railroad track with two parallel furrows for a wheeled vehicle. Some believe the goods were unloaded and carted on wagons down the diolkos to another ship. Others believe, entire ships were perched on a wheeled vehicle and moved to the sea on the other side. This saved sailors time, and it was safer than going around the Peloponnese, a peninsula in southern Greece known for violent storms.

THEN and NOW

For thousands of years, Greek fishermen sailed in boats made of papyrus reed. In AD 1988, scholars of ancient Greece recreated such a boat, which could travel from the Greek island of Corfu to Peloponnese. Even today, Greeks are among the best seamen.

16-17

Thank you gods for everything

As in most ancient civilizations, religion was an important part in the lives of the Greek people. They had many gods and goddesses, who were often related to each other and controlled everything on earth. The Greeks built grand temples for their gods. The deity was housed inside the cella, or central hall, of the temple. The ceremonies and sacrifices were mostly held outside.

Apart from temples and shrines across Greece, a special part of the home was kept aside for prayers.

▼ The Greeks took out huge processions to offer their harvest to their gods in temples.

GREEN GEM Greece has several volcanoes. It is also prone to earthquakes. This made the people worship every aspect of Nature to try and stay safe.

SMART GREEN CIVILIZATIONS

The Greeks believed the gods communicated with them through oracles, or priests who heard the divine message about the future, and communicated it to the people. Even a woman could be an oracle. The oracles stayed at particular shrines. The most respected was the oracle of Apollo at Delphi. The oracle of Zeus at Dodona was also famous. Greek mythology had a place for each of the gods and goddesses. Priests were among the most important people in Greek society.

The Greeks also believed that the way they behaved on earth helped the gods decide how to treat them after death. The gods could get angry and punish people.

▲ Apollo is a famous Greek god of music, health, and archery.

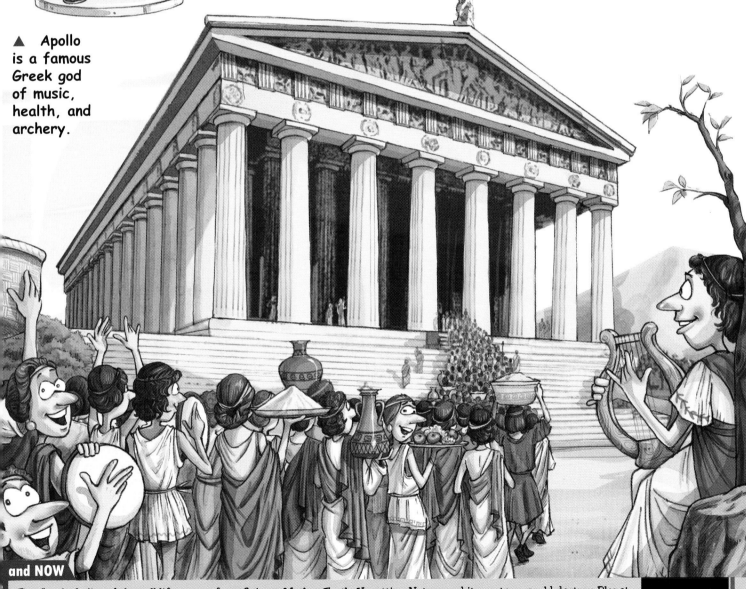

and NOW
The Greeks believed that all life sprang from Gaia, or Mother Earth. Upsetting Nature and its systems could destroy Planet Earth. People today are gradually realizing this truth.

Fitness freaks

"Gosh, did you people work out with these?"

"How do you think the Greeks grew so strong and beautiful?"

The Greeks were particular about staying healthy. Men and boys spent hours playing sports, so that they could stay fit. They believed that the gods liked those who were strong and won in sports. They participated in wrestling matches and other sports events. Boys learnt to wrestle, swim, and ride. They also learnt archery. They were excellent gymnasts. Some Greeks even did gymnastic vaults on bulls.

Most cities had a gymnasium, where people would exercise, wrestle or box. This was a courtyard lined by baths and changing rooms.

To encourage men to play, several city states organized games for athletes from different city states. Corinth held the Isthmian Games. The Olympic Games were held in Olympia to honour Zeus and his wife Hera. The first Olympic Games were held in 776 BC.

Games and wars were closely connected, as athletic men made better soldiers, who could march distances, row ships, wear heavy armour, and fight better.
Unfit soldiers, on the other hand, would make the army weak.

◄ Winners at the Olympics were given a crown of olive leaves.

▼ Chariot racing was an event at the Olympics. The winner had to cross the finish line first without crashing. With so many carts on the track, there were crashes and people would even get killed.

The health-conscious Greeks were careful about what they ate and drank. They exercised regularly and gave a lot of importance to rest and sleep.

SMART GREEN CIVILIZATIONS

The marathon race that is run today is named after Pheidippides, a Greek messenger who was sent from Marathon to Athens to tell the people that the Marathon army had defeated the Persians. In 490 BC, Pheidippides ran 240 kilometres without pausing, but he was so exhausted that he could just announce, 'We have won,' before he collapsed and died.

▲ Wrestling, running, and long jump were among the popular sports. The athletes wore no clothes!

and NOW

THEN

Some sports that were played by the Greeks like athletics, javelin, and discus throw, are still practised in the modern Olympics.

20-21

Ancient scientists

We have to thank the Greeks for many of the things we still use. They used a wide range of technology, from levers, calipers, and wind vanes to sundials.

People have been interested in numbers for thousands of years, but the Greeks were the first to study it with theories that could be proved right as far back as the sixth century BC.

Since the Greeks were so particular about health, they studied medicine with interest. Hippocrates is known as the father of medicine. He was one of the earliest doctors who tried to understand why a disease affected a person instead of treating the symptoms. Pythagoras is believed to have known that mosquitoes caused malaria. He halted the spread of malaria by pumping water out of a swamp, to stop mosquitoes from breeding.

The first Greek medical school was in Cnidus. Students were taught anatomy, or the human body, by Alcmaeon, who wrote the first known book on anatomy.

Gosh, this mosquito bite is painful!

Do you know, we had understood that mosquitoes caused malaria and took steps to control it?

▼ The Greeks studied astronomy keenly. Around 80 BC, they developed the Antikythera, a device that could calculate the movements of planets. In about 200 BC, they built an astrolabe, an instrument to calculate the distance of planets.

The Greeks knew how to use renewable energy. The Tower of Winds in Athens had a statue of Triton, which was really a wind vane to show which way the wind was blowing. It also had a water clock..

SMART GREEN CIVILIZATIONS

▲ The ancient Greeks developed many of the principles and forms of the sundial, a device to measure time by the position of the sun.

Apart from studying medicine, Hippocrates wrote on how doctors should behave. He opened a medical school at Cos. Even today, parts of his writing are read out by students when they become doctors. This is called the Hippocratic Oath.

Physicists like Ctesibius of Alexandria studied pneumatics and designed environment-friendly air and water pumps that could play a water organ, operate a pump, and even a cannon. The Greeks also built the first known water mill.

Philo wrote in 'Pneumatics' that the Greeks used water power to drive a water mill that could do work like grinding grain. They also knew how to search for precious metals like silver, which they mined from underground.

◀ Greek doctors used a process called bloodletting to treat diseases. They did this by making a cut on the patient's arm until blood ran out.

THEN and NOW
The Greeks devised a pump that could be used to douse fires with a jet of water. The idea was forgotten some centuries later. Recently, we began fighting fires with water jets once again.

Art and architecture

That's the temple of Zeus on Mount Olympus. It was built for me during 470-456 BC.

It's bigger than any temple I've seen!

The ancient Greeks are still remembered for their paintings, statues, architecture, pottery, theatre, and music. Their earliest buildings were made of wood or clay. The Greeks were expert builders. By 600 BC, they had learnt to replace parts of a wooden structure with stone without damaging the building. The columns and other parts of the Temple of Hera at Olympia were changed to stone.

▲ Greek builders used a large number of tools made of iron, bronze, and wood to make buildings.

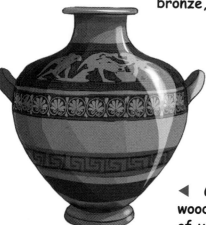

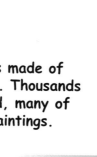

◄ Greek pottery was made of wood, clay or ceramic. Thousands of vases have survived, many of them with beautiful paintings.

 Greek buildings were built with locally available natural material like marble, mud or terracotta, wood, limestone, and bricks.

SMART GREEN CIVILIZATIONS

The temples are the best known Greek buildings. Some had a porch that opened on to a yard. The tholos were round buildings. The Greeks were sociable people, and they built different types of public buildings. These were shops, or agoras, fountain houses to fill water from, and government houses and theatres with semi-circular steps for seating.

The Greeks built houses to live in, temples, gymnasiums, open-air theatres, and tombs. Some of the best buildings were built between 550 BC and 400 BC. In structures like the temple of Zeus, they used local limestone, so that they did not have to go far to get the material. To make the building look better, they coated the stone with stucco, a kind of plaster.

The Greeks combined their expertise in different arts with their observation of nature. They worked metals like gold and copper. Sculptures could be tiny or larger than life. Statues like the Snake Goddess show the Greeks' respect for animals.

▼ The architecture of ancient Greece is famous for the most decorative columns.

◄► Greek builders took great care to draw a plan, according to which their buildings were built.

THEN and NOW

The Greeks limewashed their houses, which killed germs and pests and brightened the walls. Even today, many Greek homes use limewash rather than paints, which may have harmful chemicals.

The battles they fought

The Greeks were a strong military power in the ancient world. City states like Sparta and Athens were powerful military forces that sometimes fought each other, as in the Peloponnesian War. Some wars were fought by one or more city states against other people. In the Battle of Marathon in 490 BC, soldiers from Plataea joined the Athenians to fight the Persians. Wars like the Battle of Salamis against the Persians led by their king Darius were fought by the navies. Foot soldiers were an important part of the Greek military. The Greeks marched in tight formations, or phalanxes.

The Greeks designed their cities to keep them secure from enemies. The walls had high watch towers. The towers had shutters from where guards could keep watch and shoot the approaching enemy down with arrows.

The most famous battle story from Greece is the tale of the Trojan Horse. Even after ten years of surrounding the city of Troy, the Greeks could not capture it.

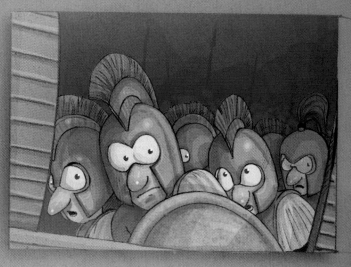

▲ While the Trojans slept at night, thirty Greek soldiers hidden inside the wooden horse climbed out and opened the gates of Troy!

The Greeks used renewable energy even in their military campaigns. Their ships, carrying soldiers, sailed with the help of wind energy. When they needed to speed up, ships were rowed by sailors.

SMART GREEN CIVILIZATIONS

▼ The Trojan Horse was so big, it would not go through the gate. The Trojans had to tear down a piece of the city wall to let it in!

So, they built an enormous hollow wooden horse on wheels. A group of soldiers hid in it. The horse was left at the gates of Troy. The Greeks pretended to sail away. Believing it to be a gift to Goddess Athena, the Trojans wheeled the horse in after breaking the wall to let in the horse. That night, while the Trojans slept, Greek soldiers crept out and signalled to their ships to return. The Greeks won the war and captured Troy.

THEN and NOW

War is never good for the environment. But today's wars use more machines than humans, because machines can destroy more and faster. Modern scientists are thinking of making biodegradable tanks to reduce the pollution caused by war.

Why did they disappear?

Why did this glorious civilization end?

It took many reasons and many seasons to bring on the decline of the ancient Greek civilization.

The rise and fall of civilizations are due to a patchwork of causes. The geography of Greece was one reason for its decline. What is today known as Greece consists of several pieces of land divided by water. The mainland is partitioned into smaller regions by mountains. This led to the growth of city states that were cities but were governed like little countries.

The existence of several city states was also one reason the Greeks could not unite into one strong country. Instead, people of the city states became used to believing they were separate. This also made them more competitive. As rivalry increased, so did battles between the city states. War is always expensive. Since they could not unite, others took advantage of their differences. Neighbouring kingdoms united and grew in strength. Philip of Macedonia, who had a large army, conquered Greek city states by 338 BC. By 146 BC, the Romans had conquered most of Greece.

◀ The Greeks were taking more from the land than it could bear. More land was dug up for mining, and more trees were cut down to melt metals. This made the land barren, forcing people to migrate.

GREEN GEM — As the number of cattle grew, they needed more pasture to graze in. Greece is an arid region and the loss of forests made it more barren.

SMART GREEN CIVILIZATIONS

▲ Frequent wars caused a lot of destruction and damage and led to a decline in the quality of life.

The Greeks tried to come together after a war with the Persians. They grew so strong that the Persians could not take over. But powerful city states like Athens and Sparta could not trust each other. This led to the Peloponnesian War (431–404 BC). It divided all the city states into two groups. No sooner did Sparta win this war than Thebes, its old ally, defeated it. Such wars continued and weakened Greece.

THEN and NOW

As time went by, the population grew and cities became crowded and dirty. People cut forests to grow crops and to build more houses. The land became dry and arid and could not sustain them. Sounds familiar? We are facing similar problems today!

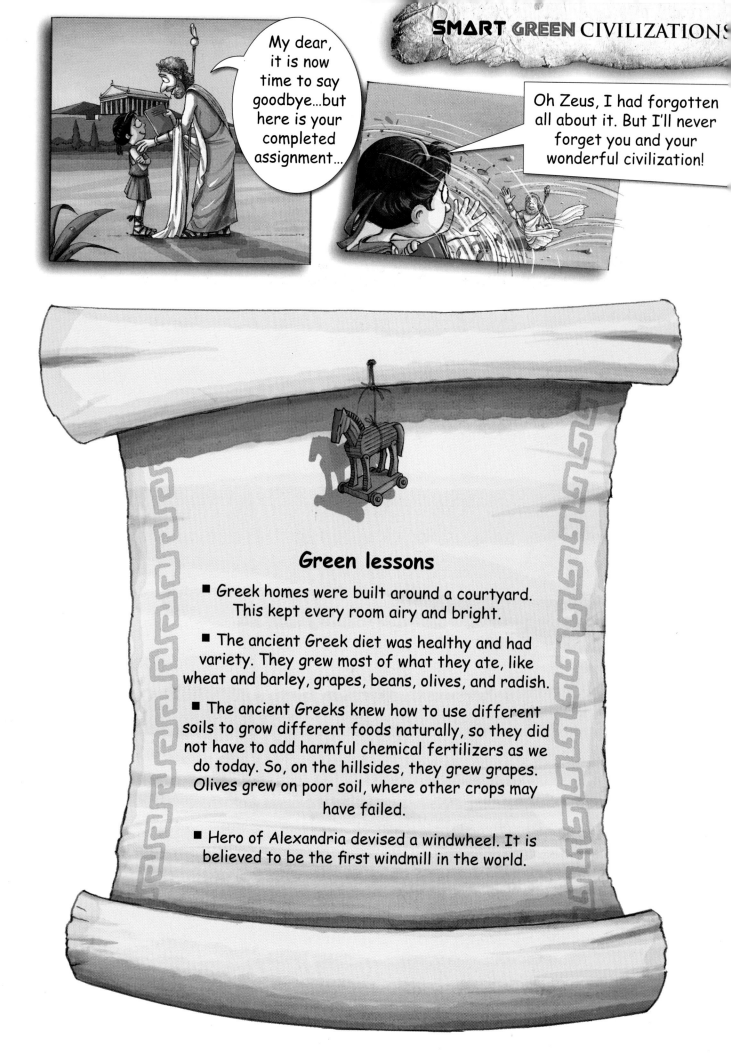